Geometry Workbook 6th to 7th Grade

(Baby Professor Learning Books)

Speedy Publishing LLC
40 E. Main St. #1156
Newark, DE 19711
www.speedypublishing.com

Triangles

Classify Triangles by their angles

(acute, right or obtuse)

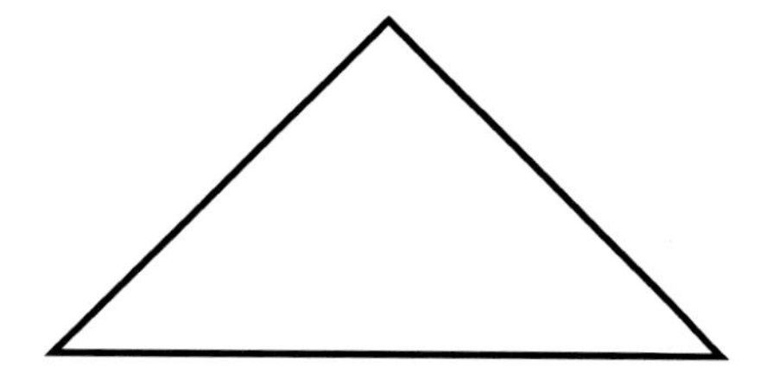

____

____

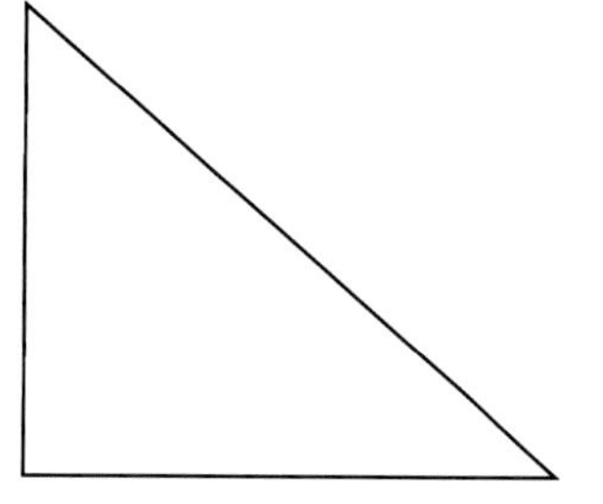

____

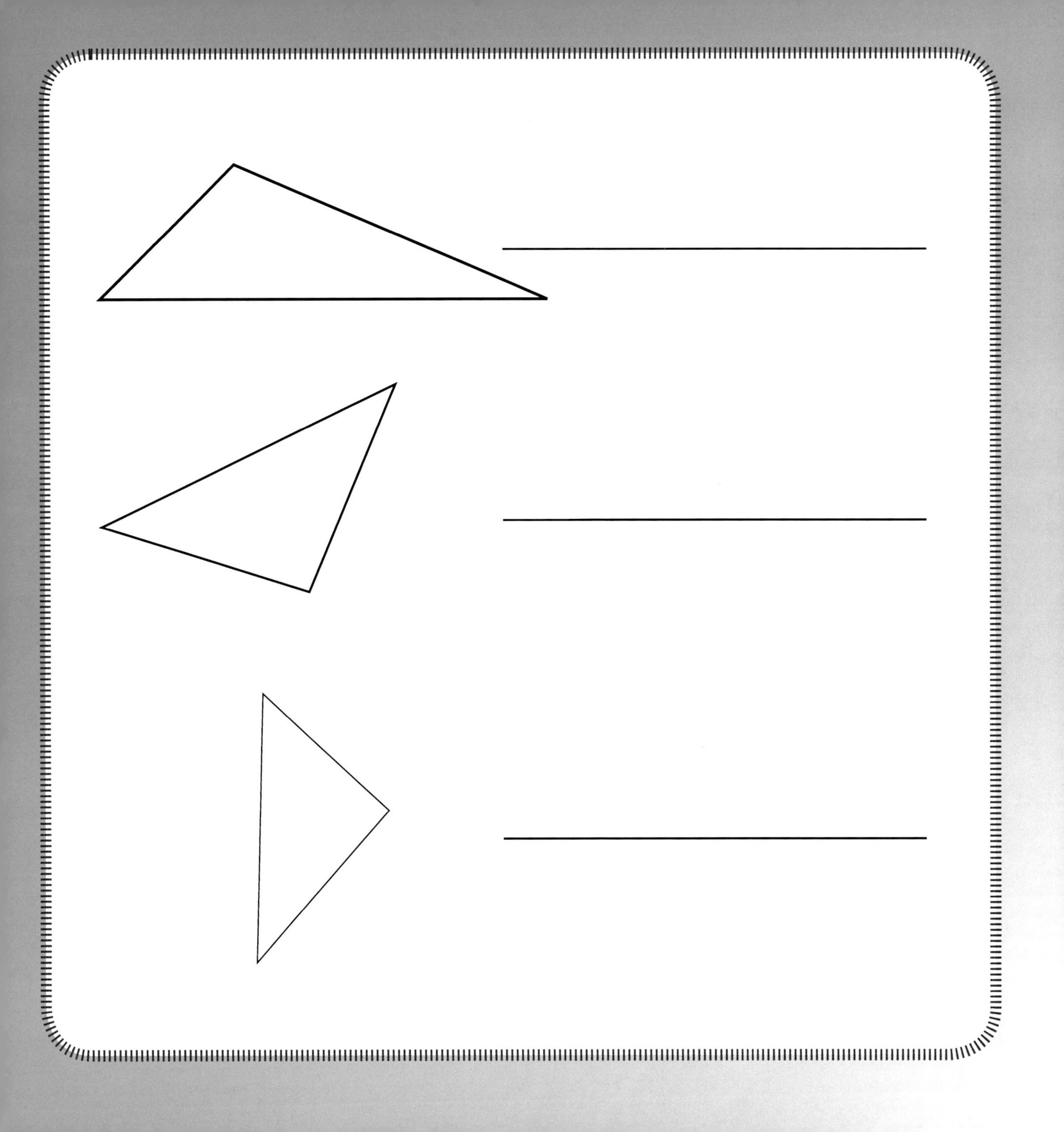

Calculate.
Show your solutions.

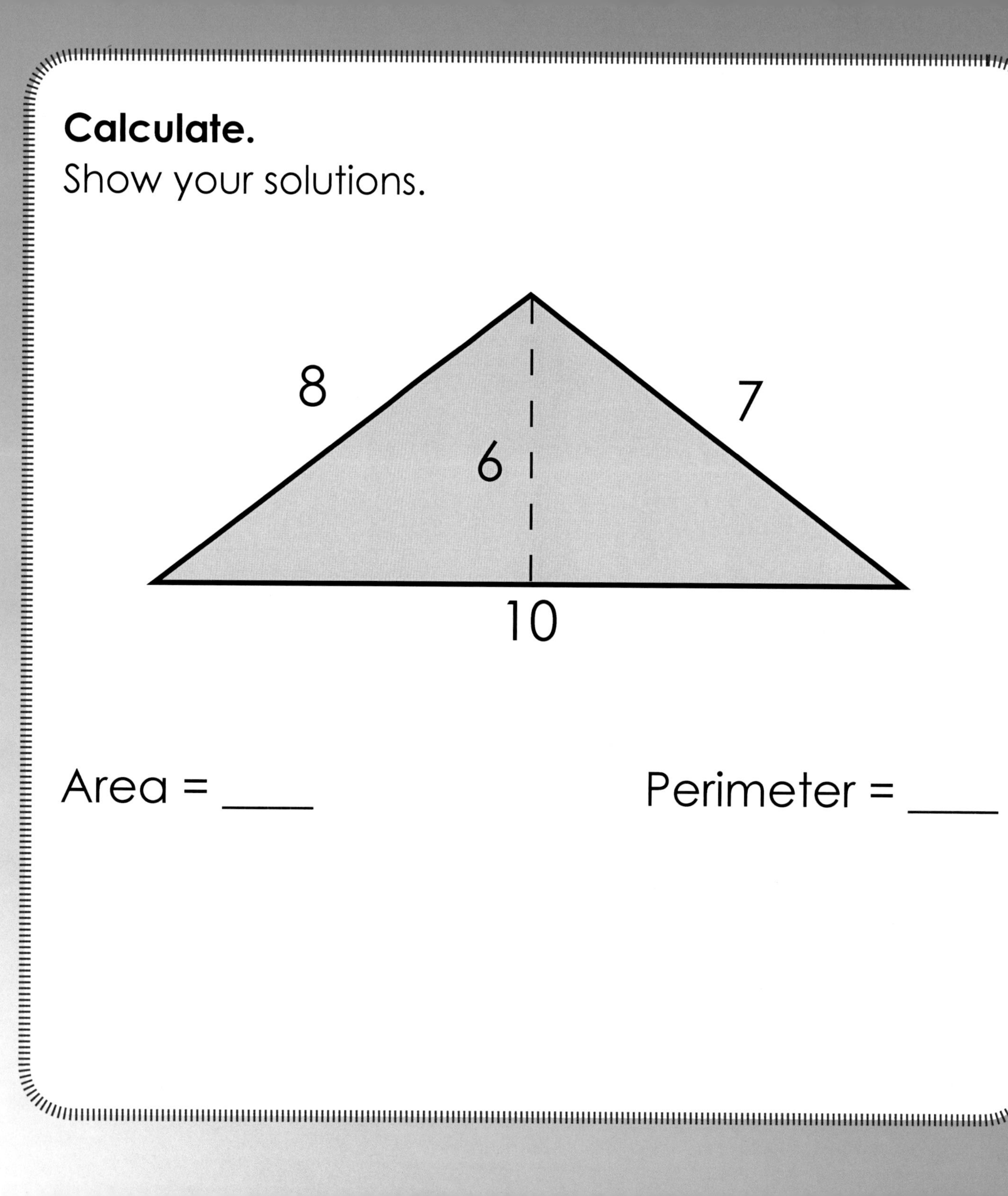

Area = ____

Perimeter = ____

Calculate.

Show your solutions.

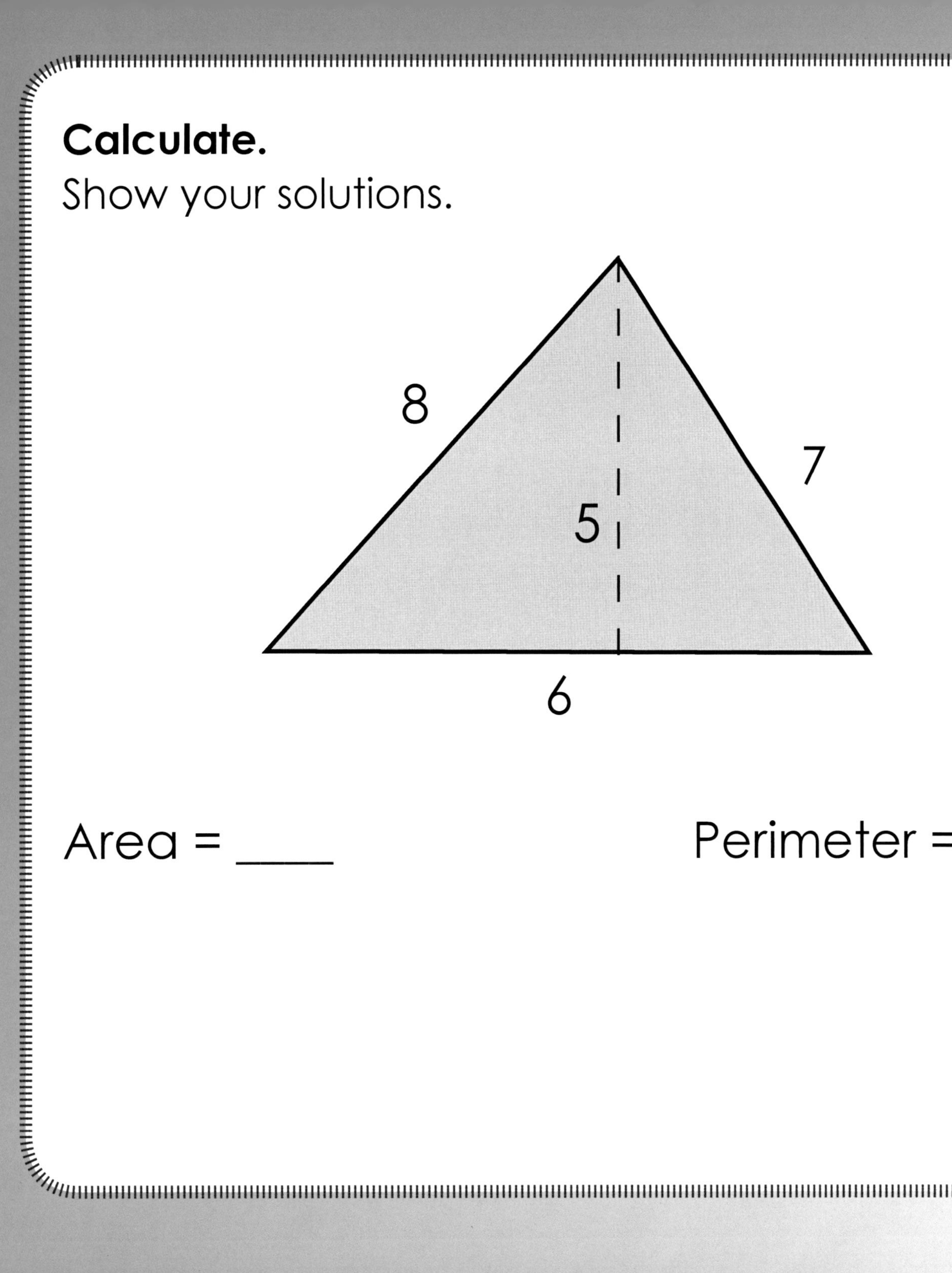

Area = ____

Perimeter = ____

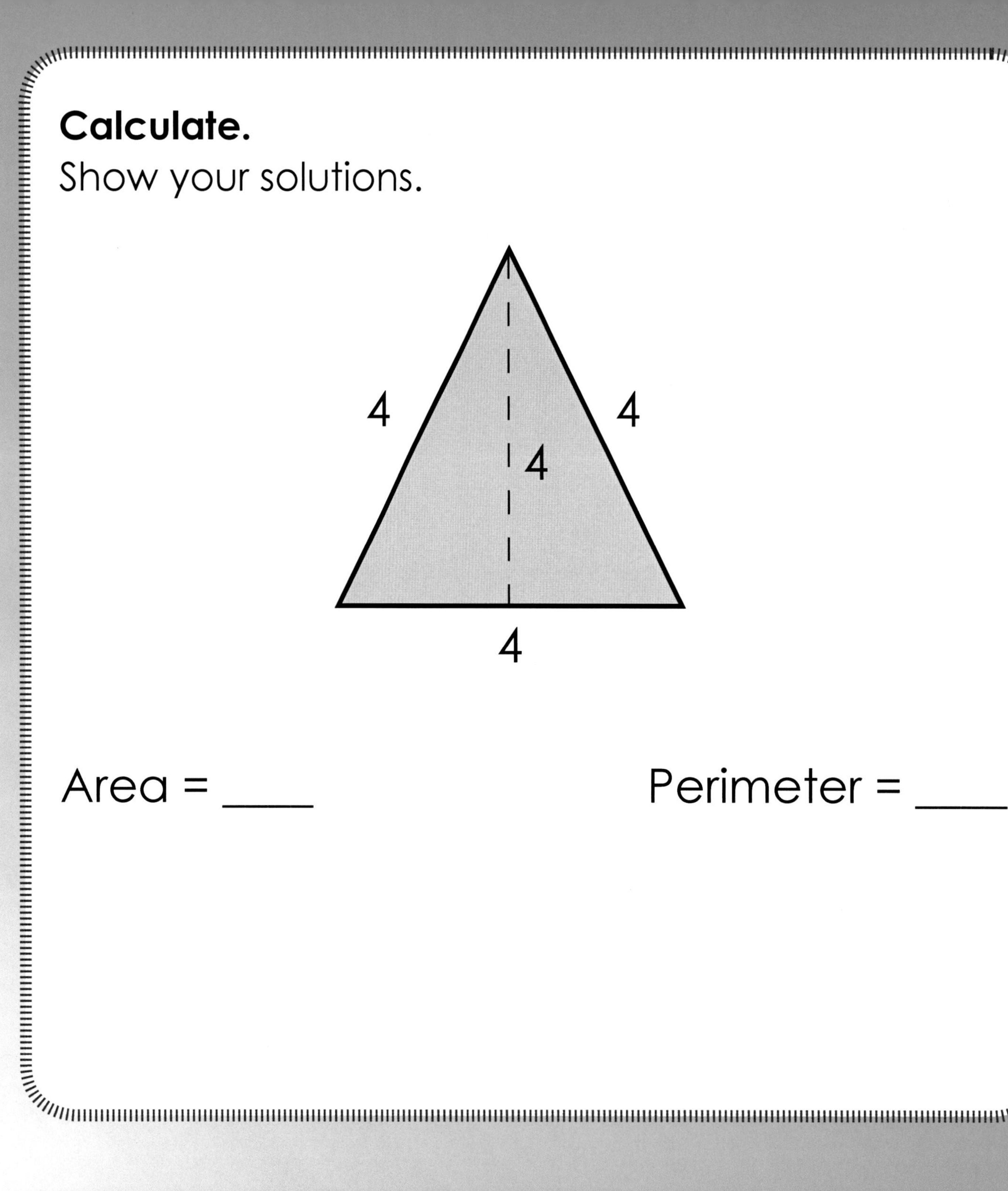
Calculate.

Show your solutions.

Area = ____

Perimeter = ____

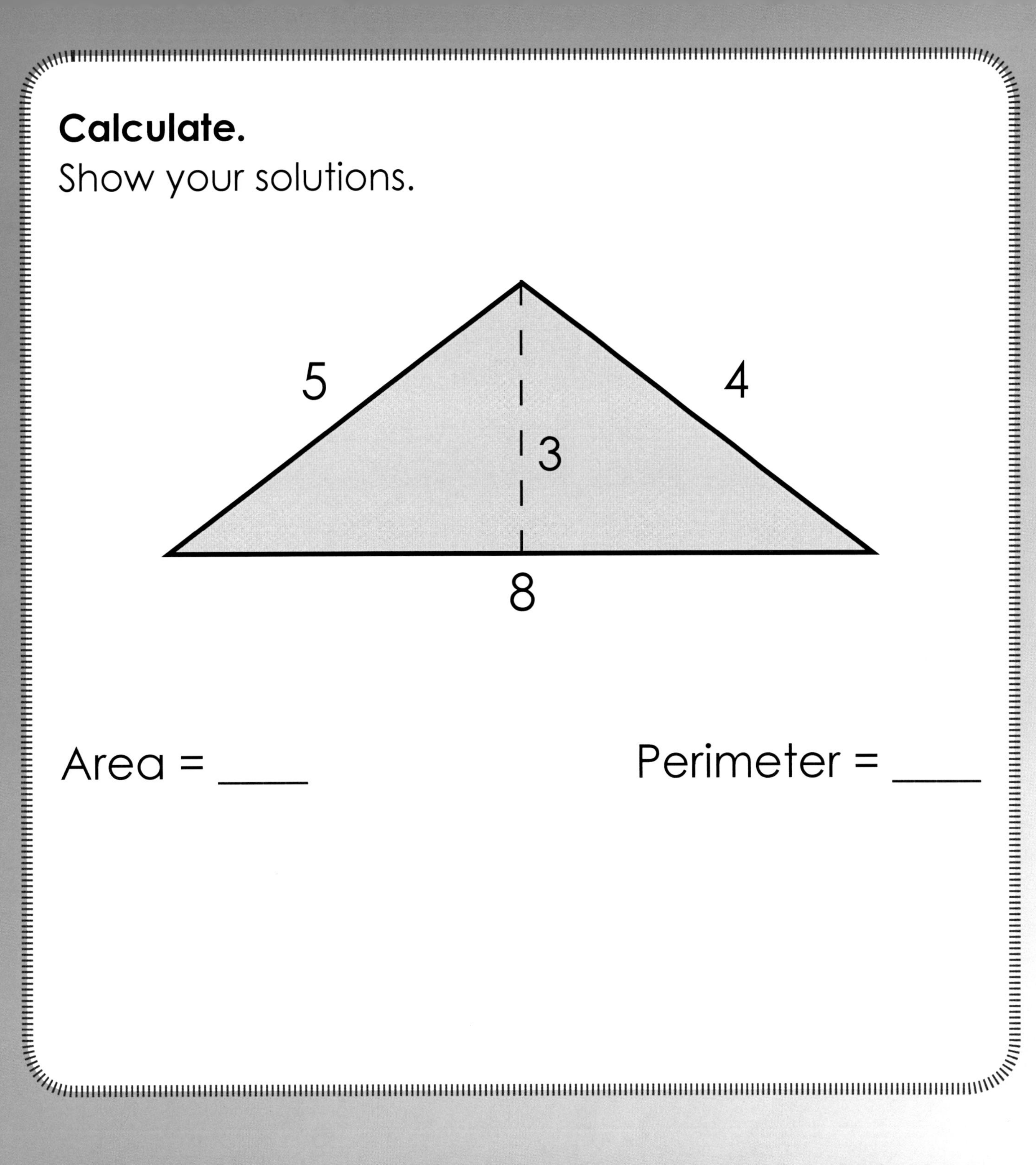
Calculate.
Show your solutions.
5
4
3
8
Area = ____
Perimeter = ____

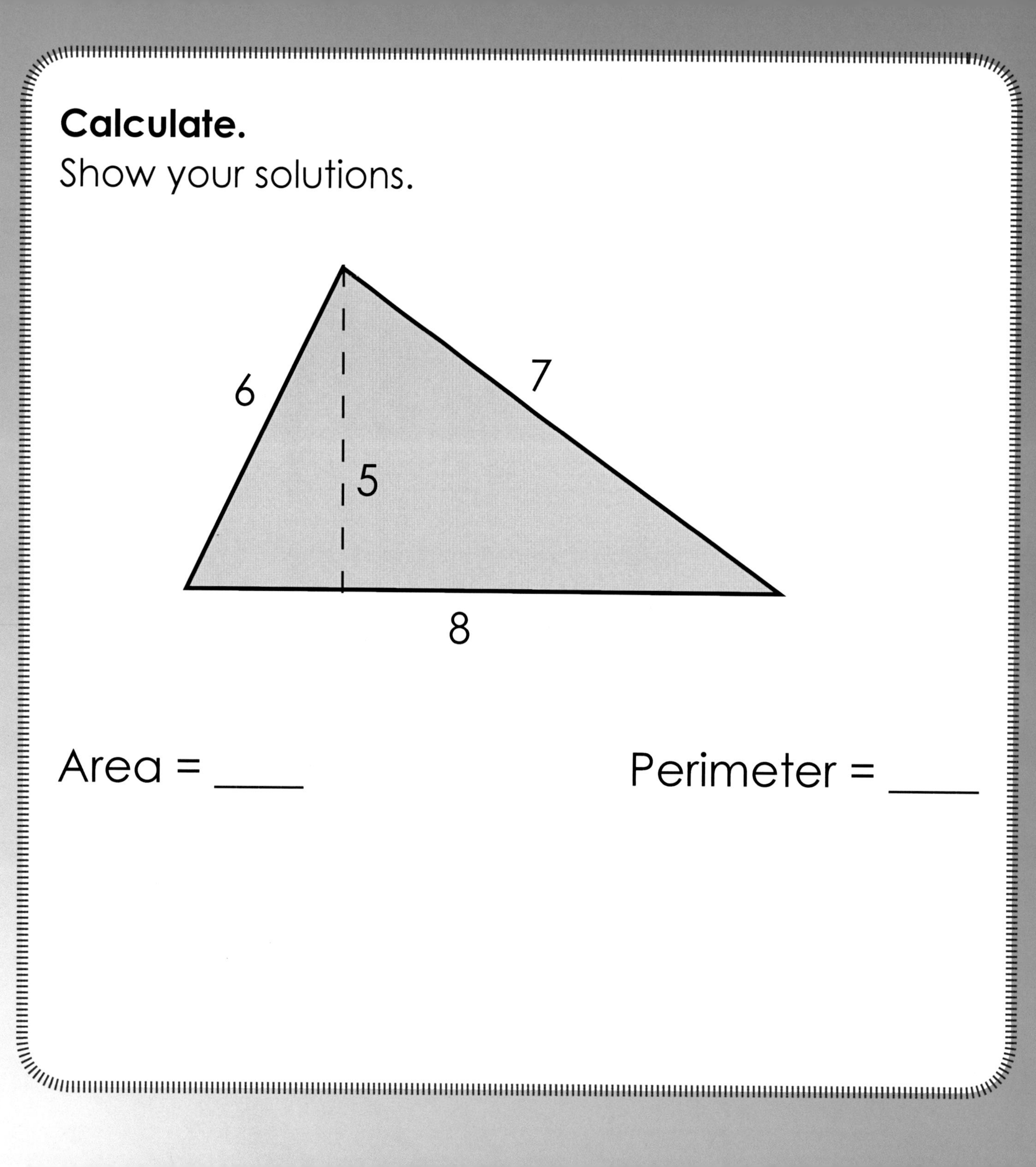
Calculate.
Show your solutions.
6
7
5
8
Area = ____
Perimeter = ____

Calculate.

Show your solutions.

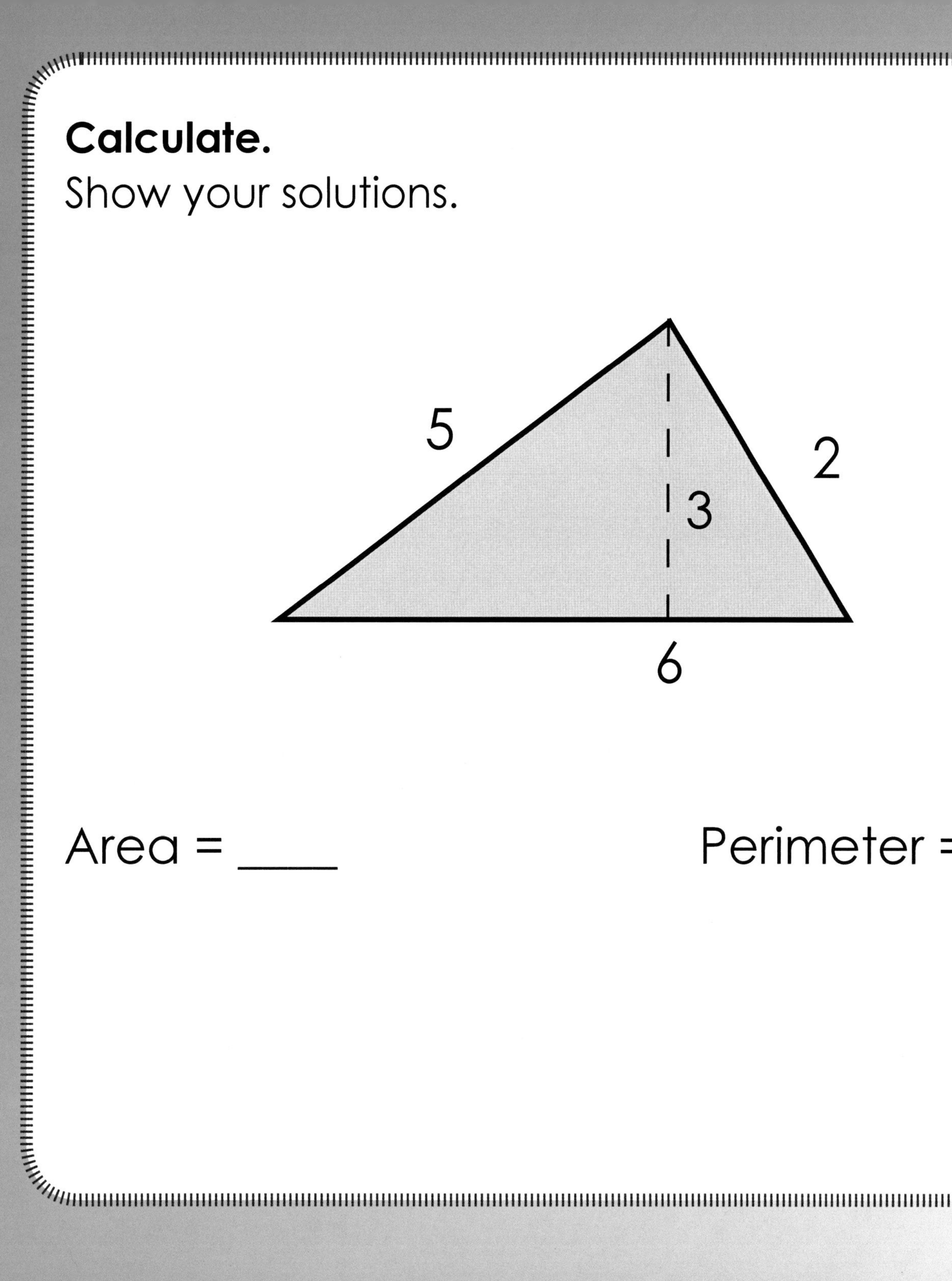

Area = ____

Perimeter = ____

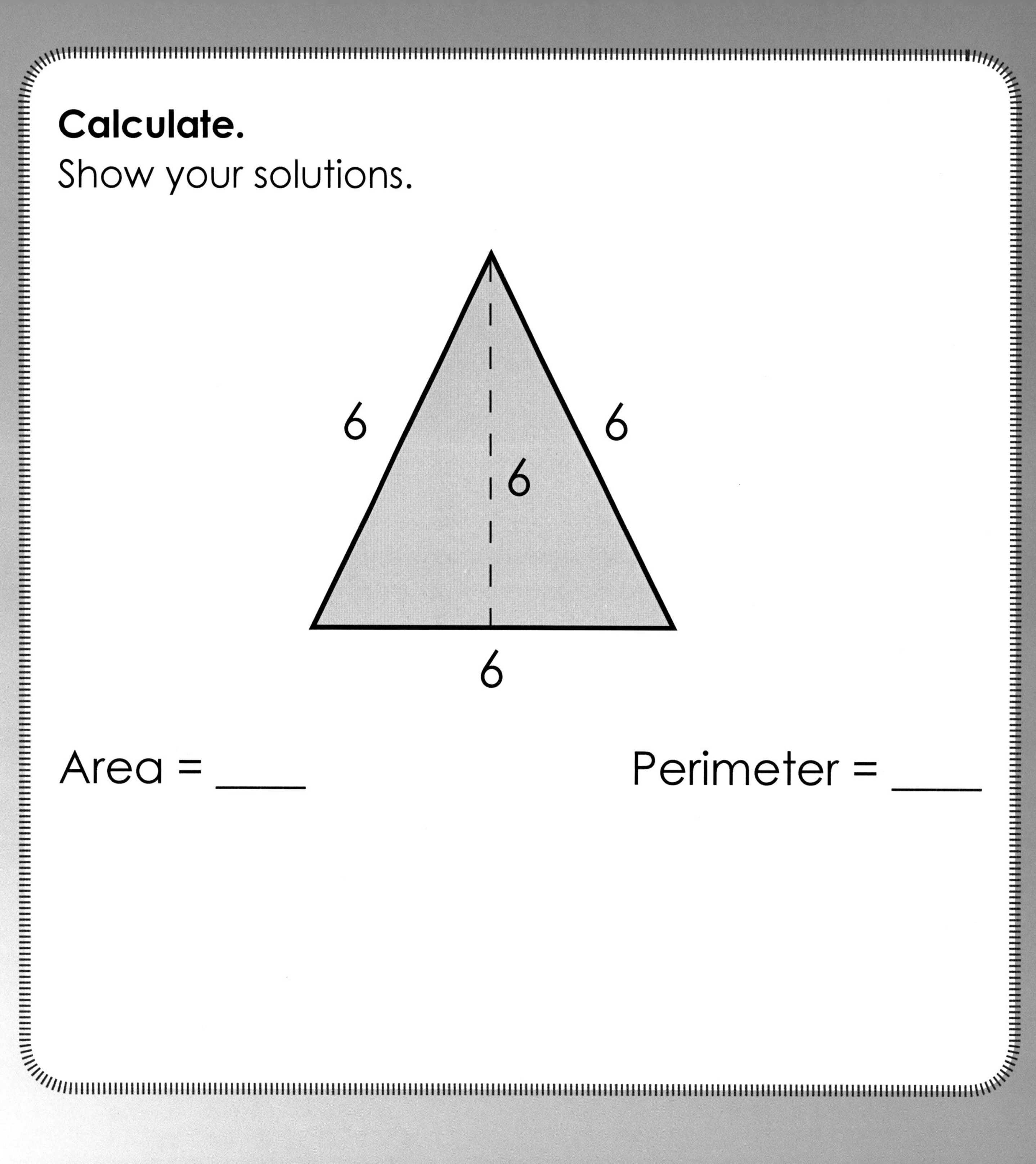
Calculate.
Show your solutions.
6
6
6
6
Area = ____
Perimeter = ____

Circles

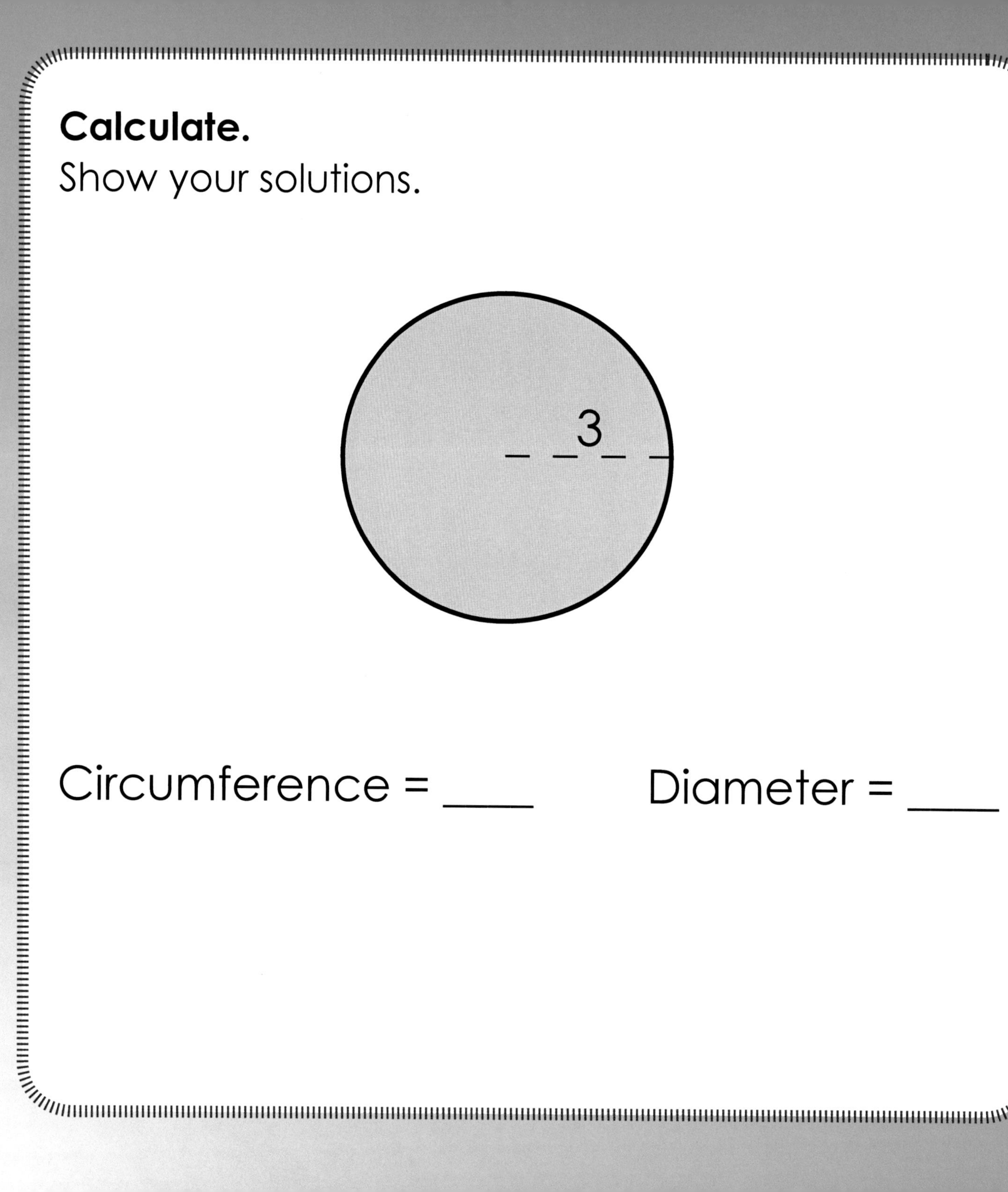

Calculate.

Show your solutions.

Circumference = ____

Diameter = ____

Calculate.

Show your solutions.

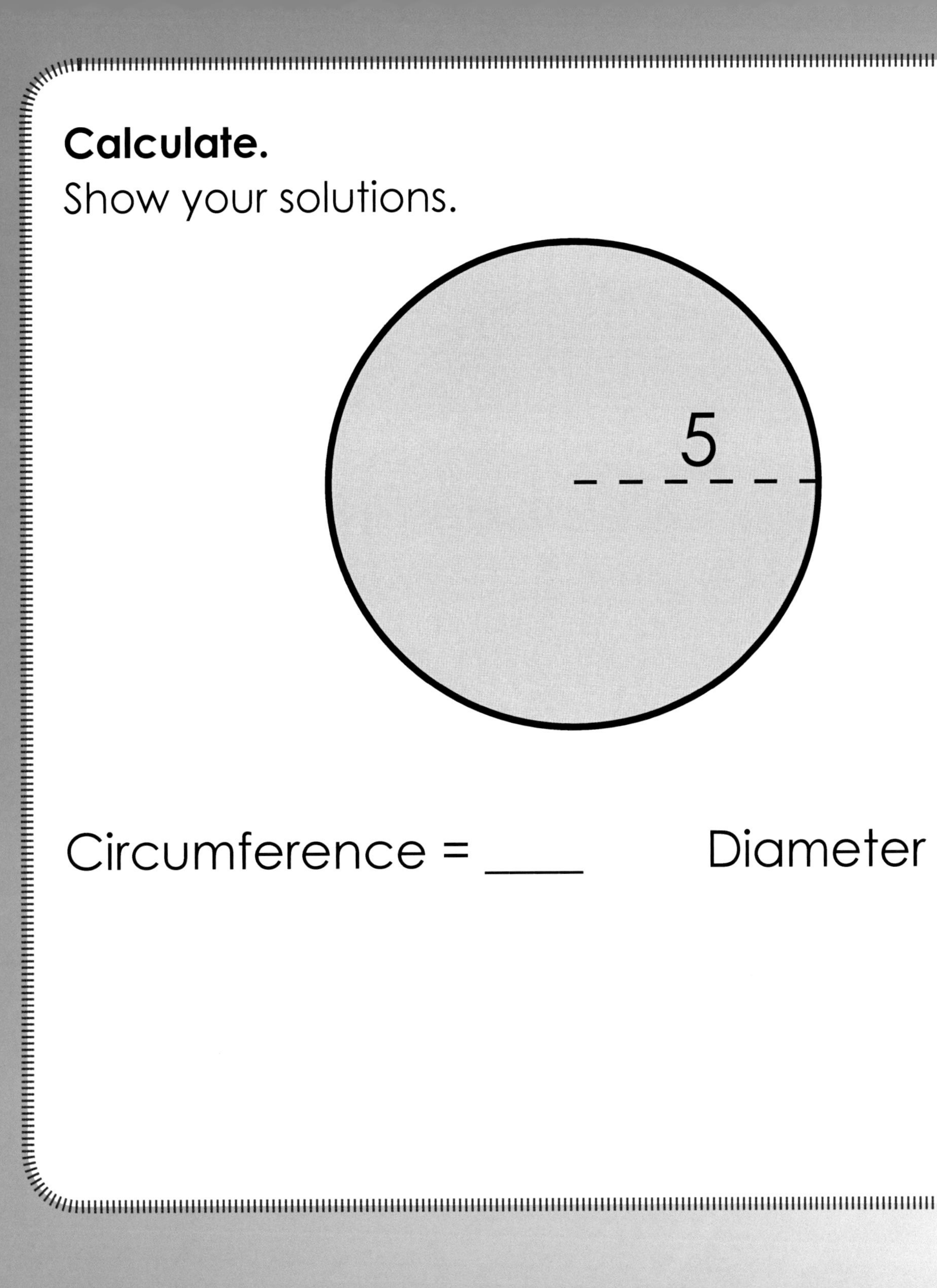

Circumference = ____   Diameter = ____

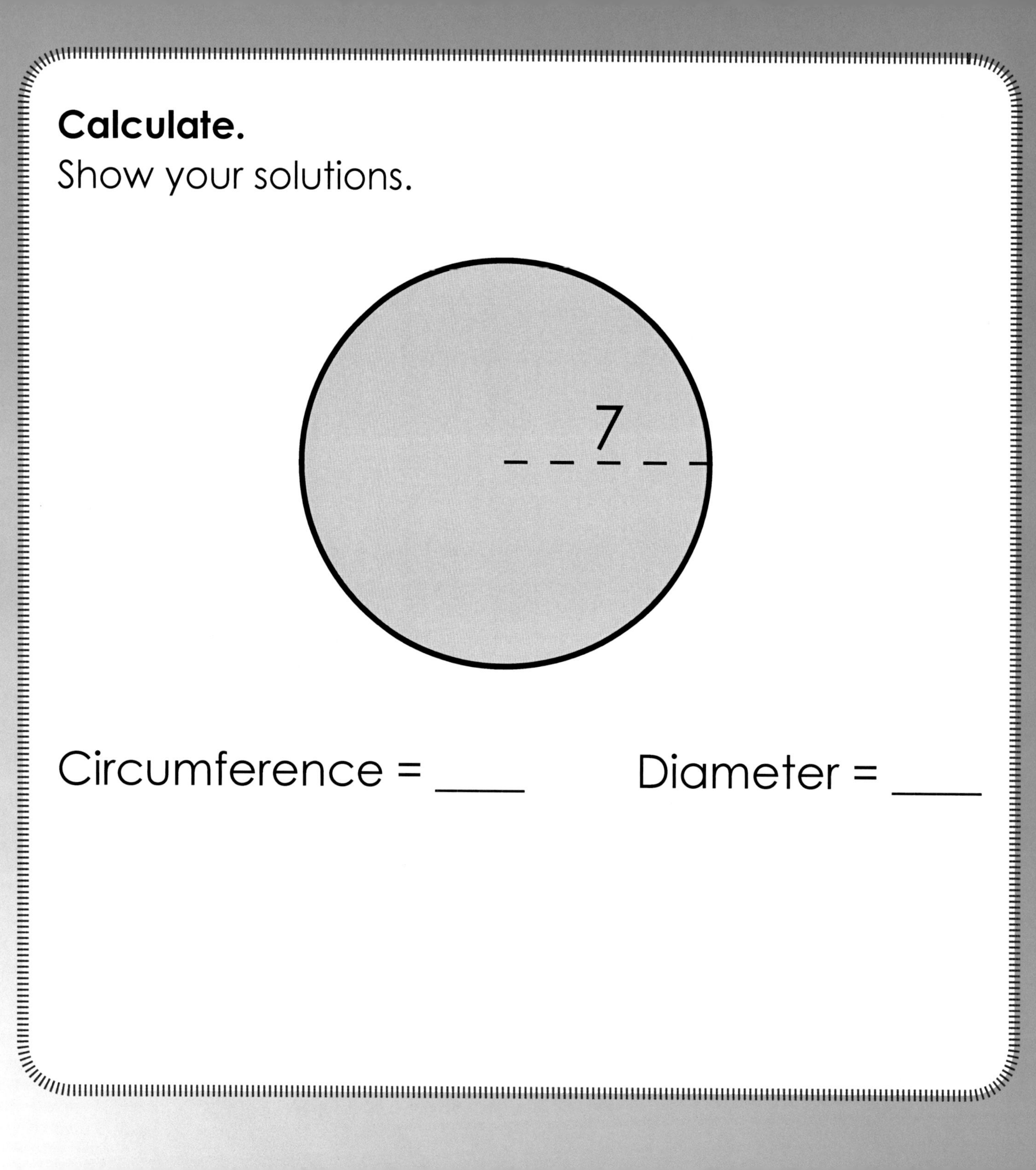

Calculate.

Show your solutions.

Circumference = ____   Diameter = ____

Calculate.

Show your solutions.

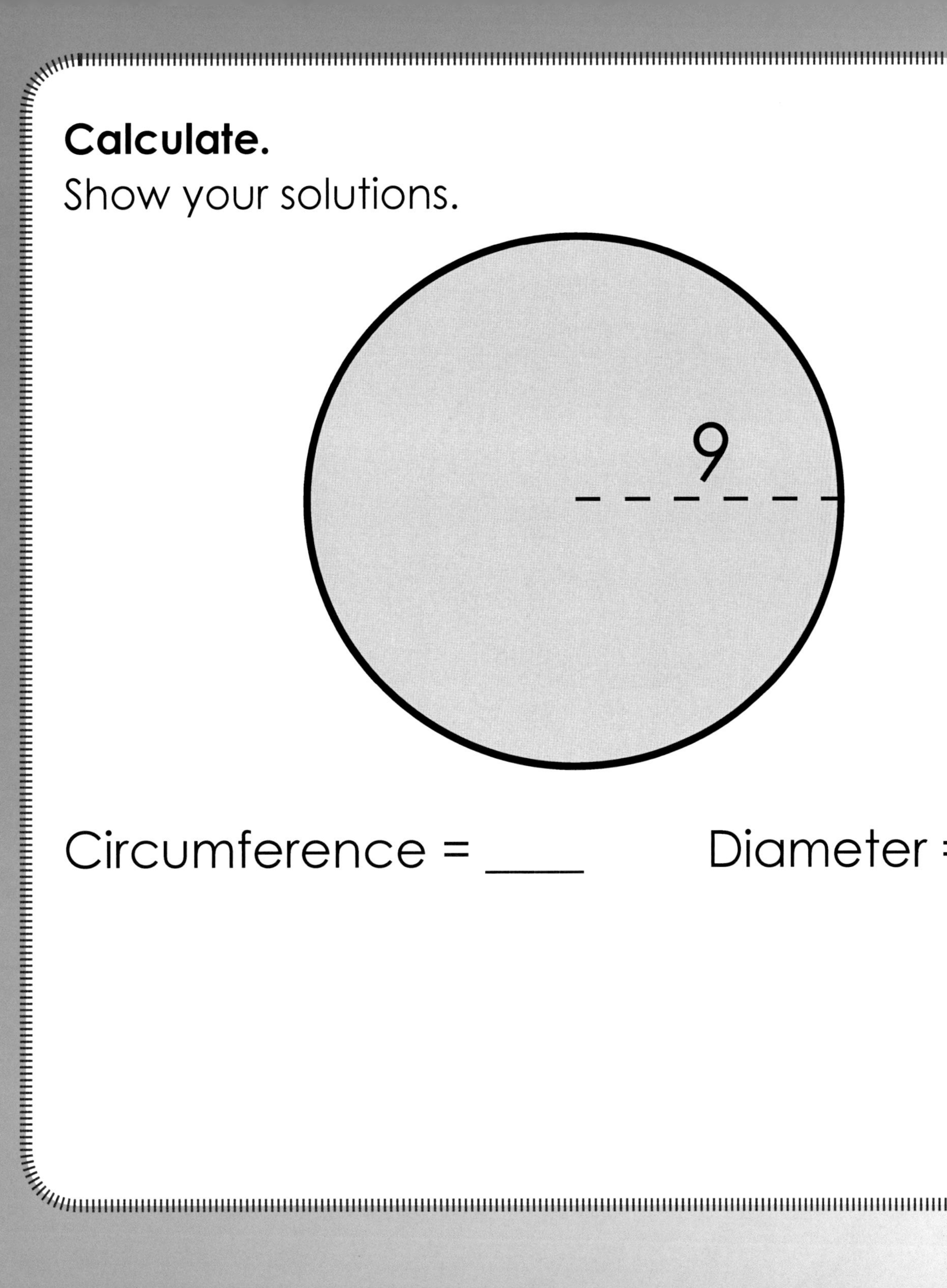

Circumference = ____ Diameter = ____

Calculate.

Show your solutions.

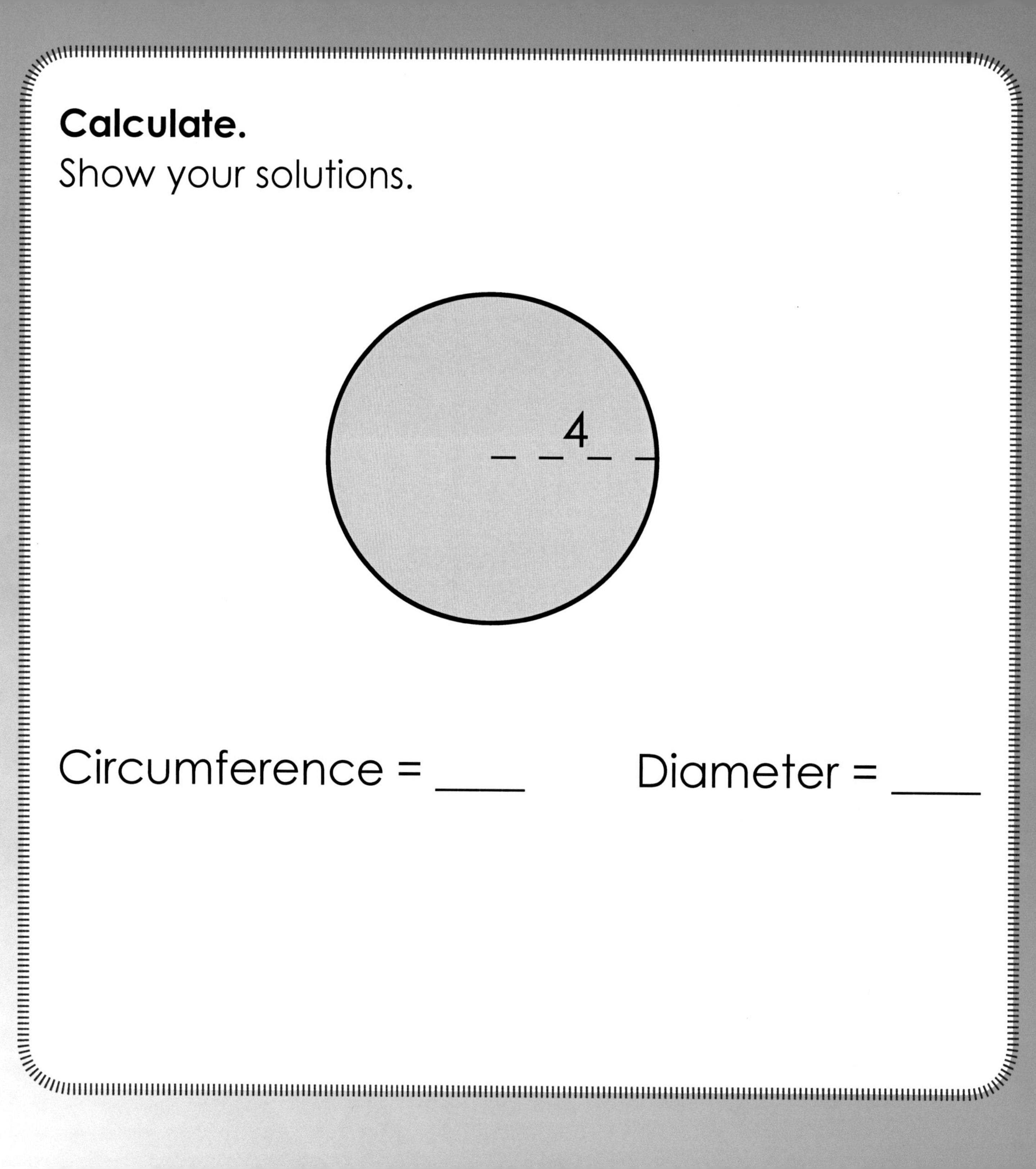

Circumference = ____ Diameter = ____

Calculate.

Show your solutions.

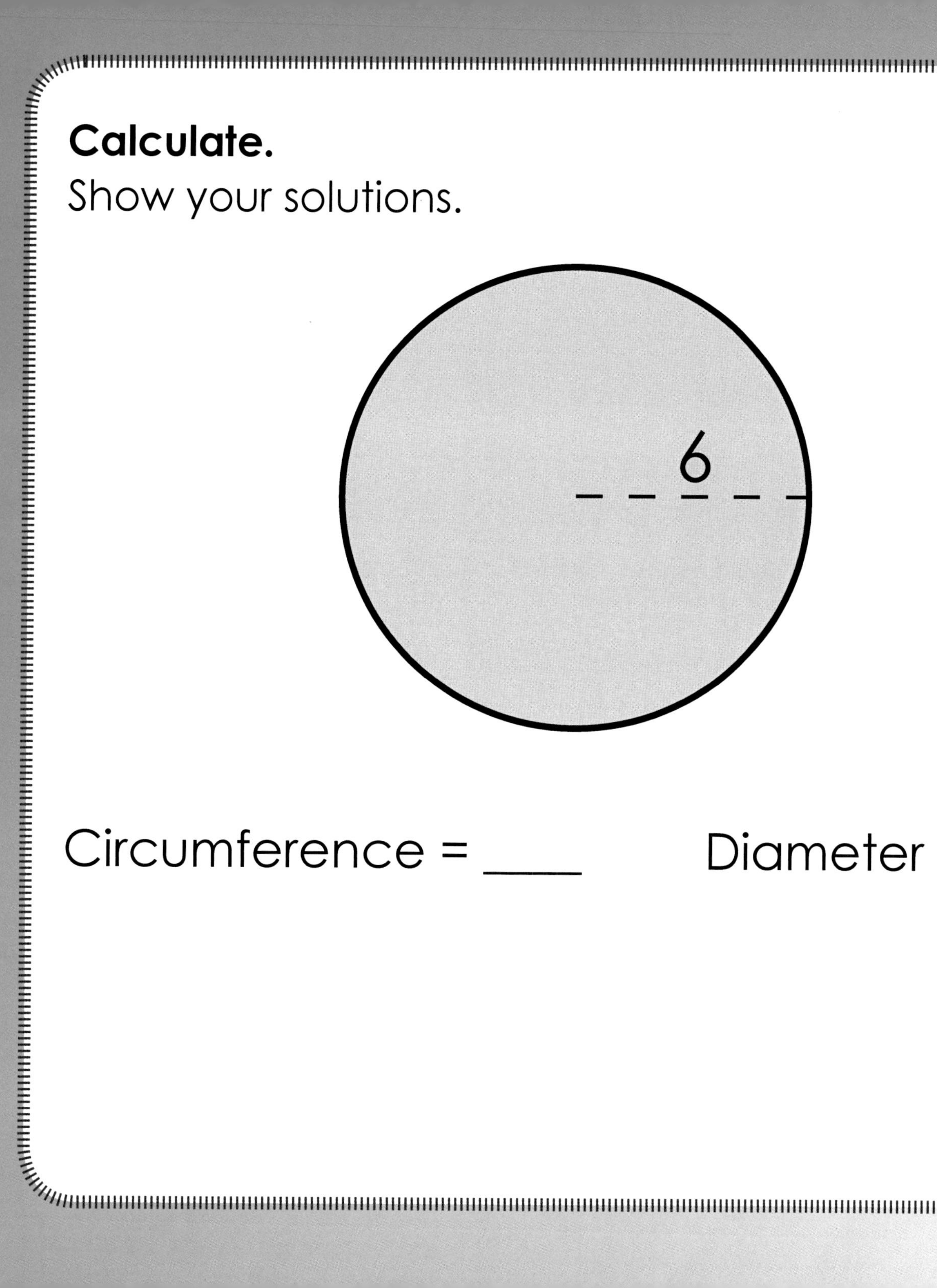

Circumference = ____   Diameter = ____

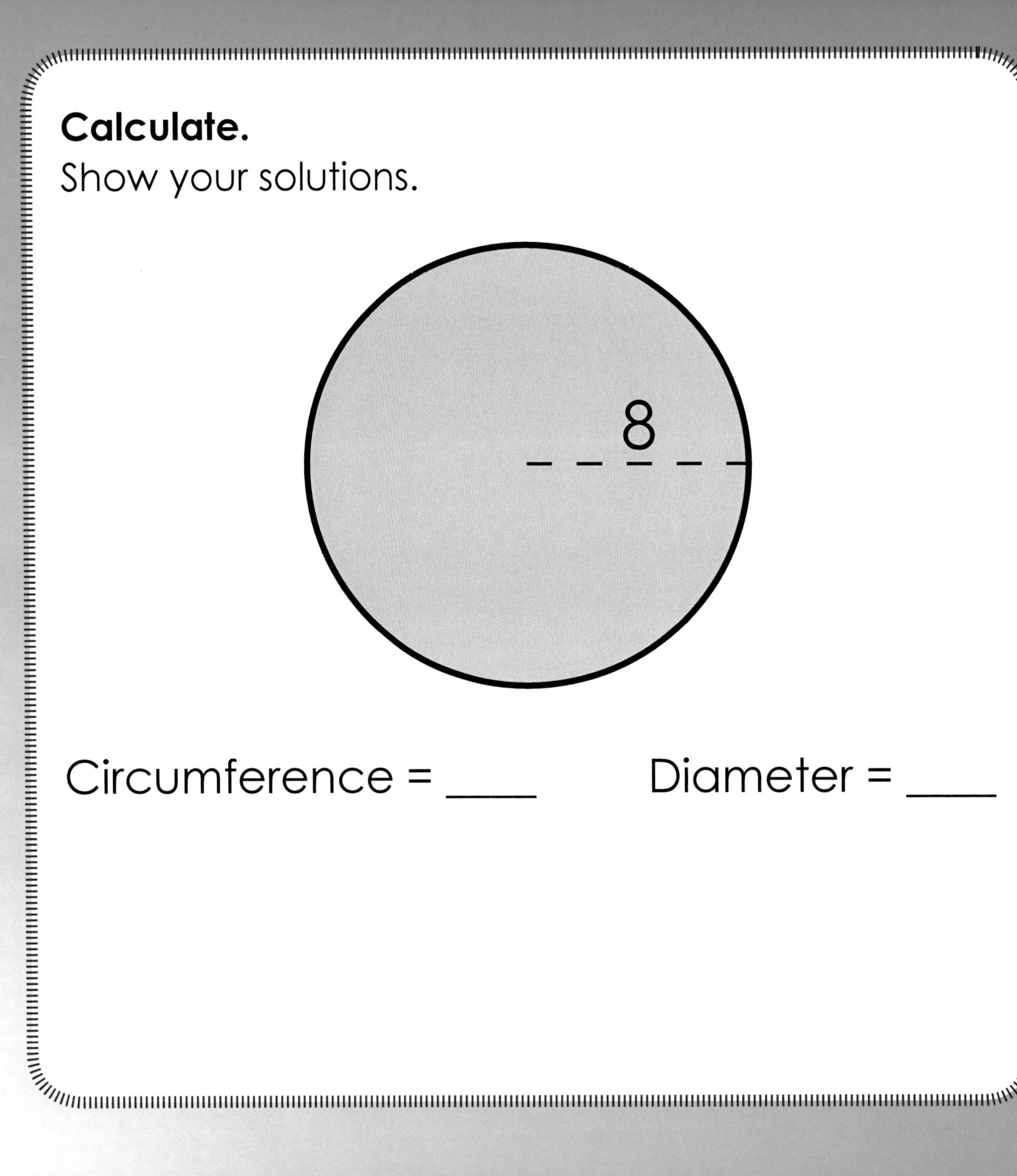

Calculate.

Show your solutions.

Circumference = ____ Diameter = ____

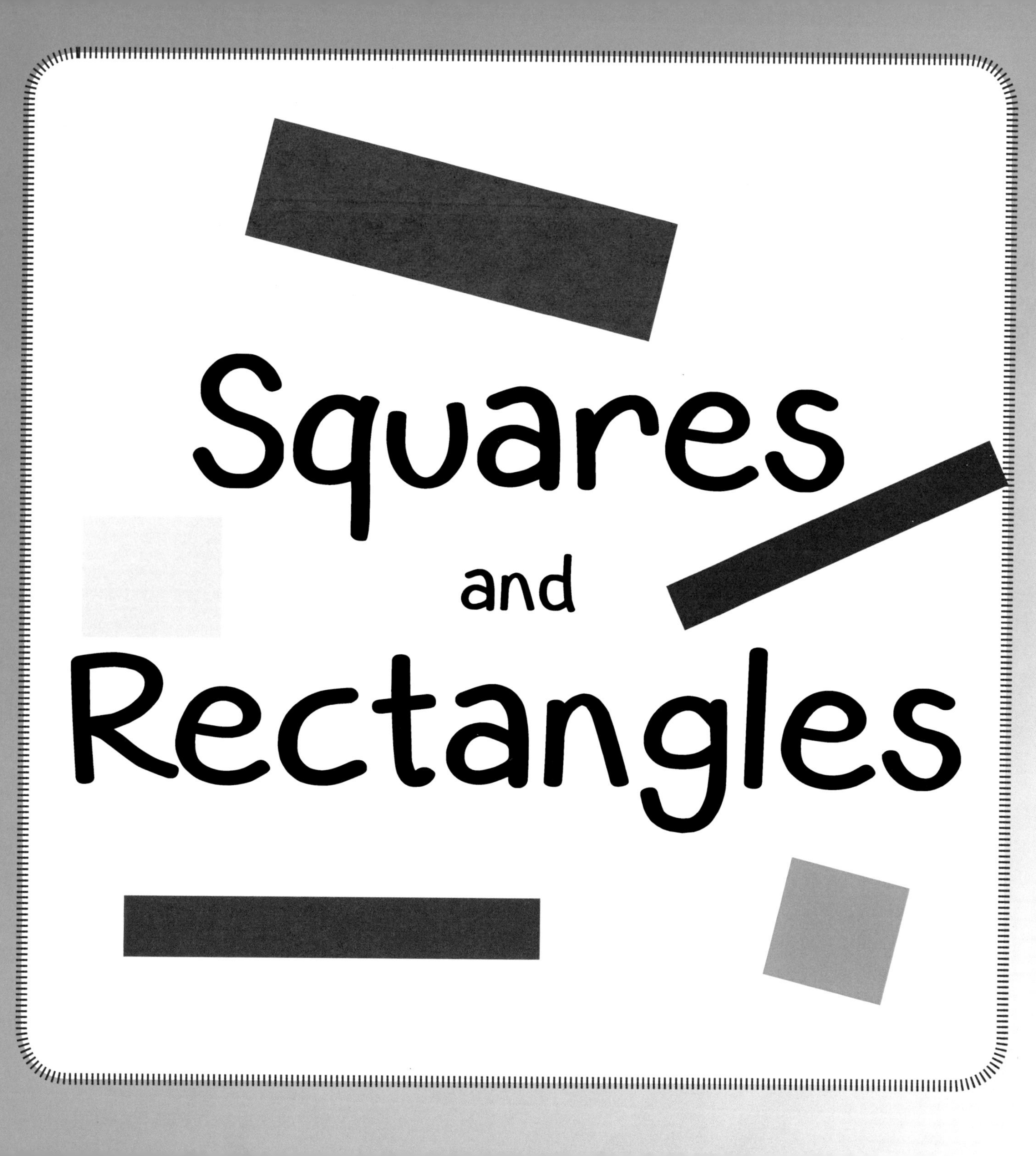
Squares
and
Rectangles

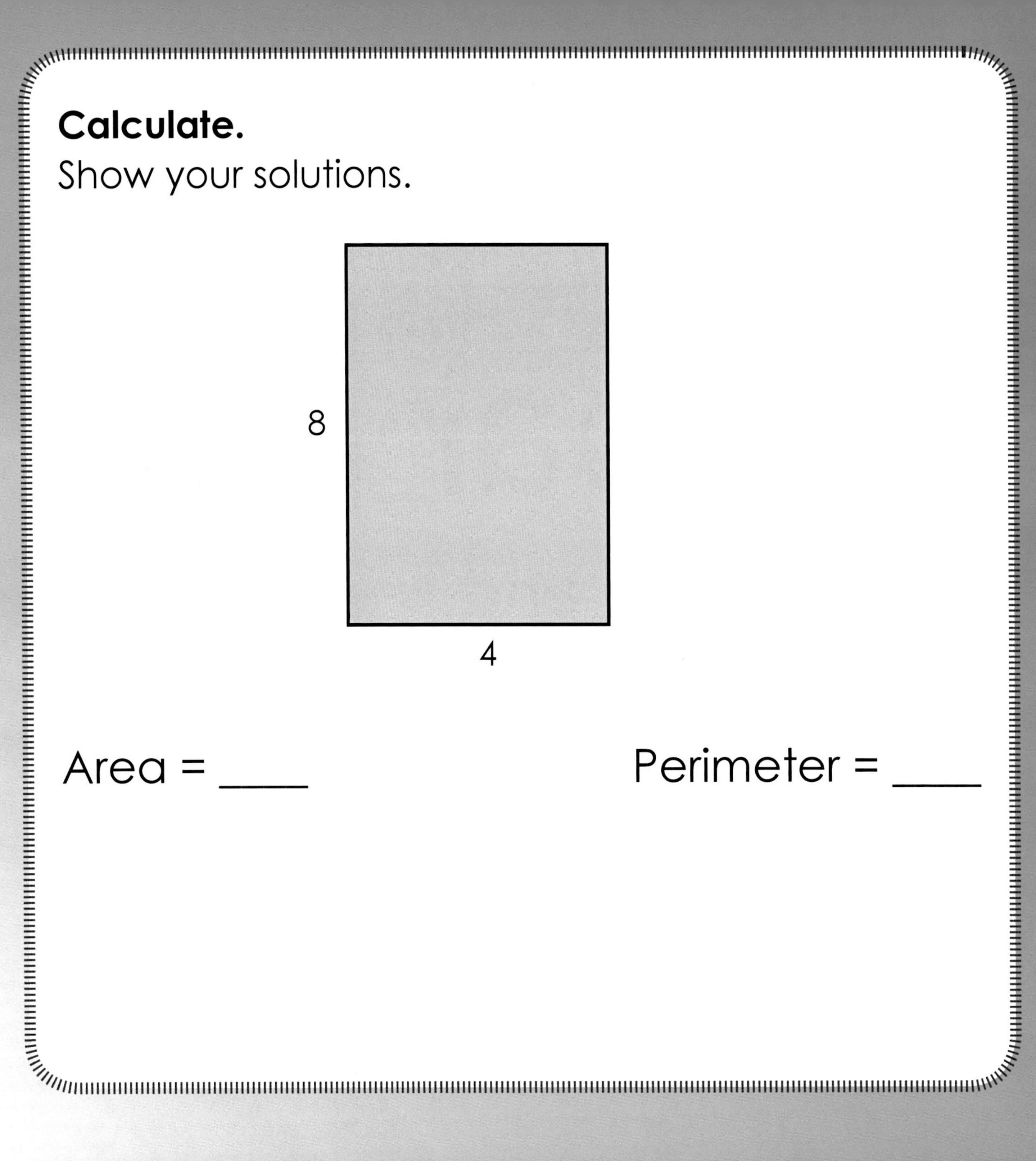

Calculate.

Show your solutions.

Area = ____

Perimeter = ____

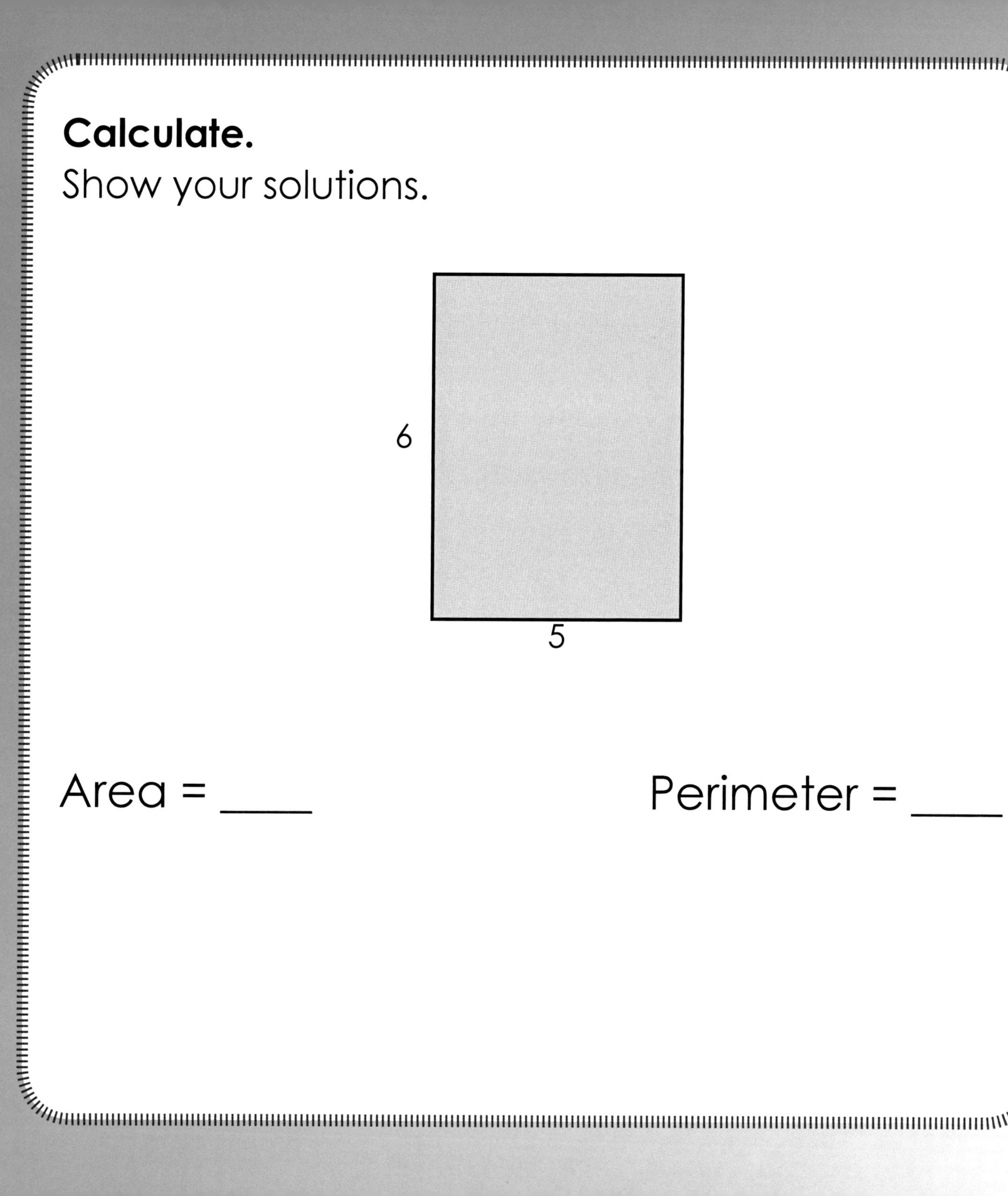
Calculate.
Show your solutions.
6
5
Area = ____
Perimeter = ____

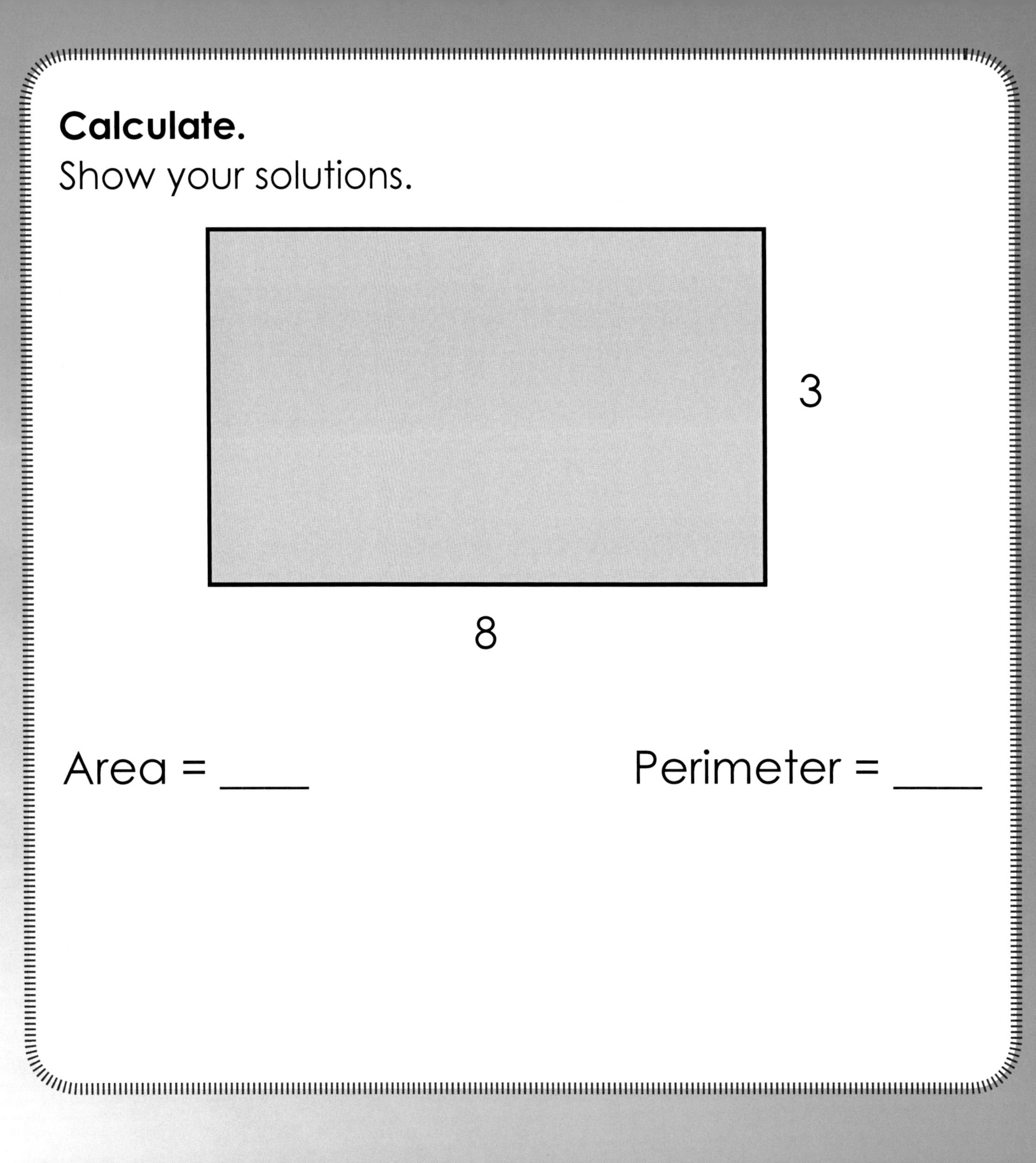

Calculate.

Show your solutions.

Area = ____

Perimeter = ____

Calculate.

Show your solutions.

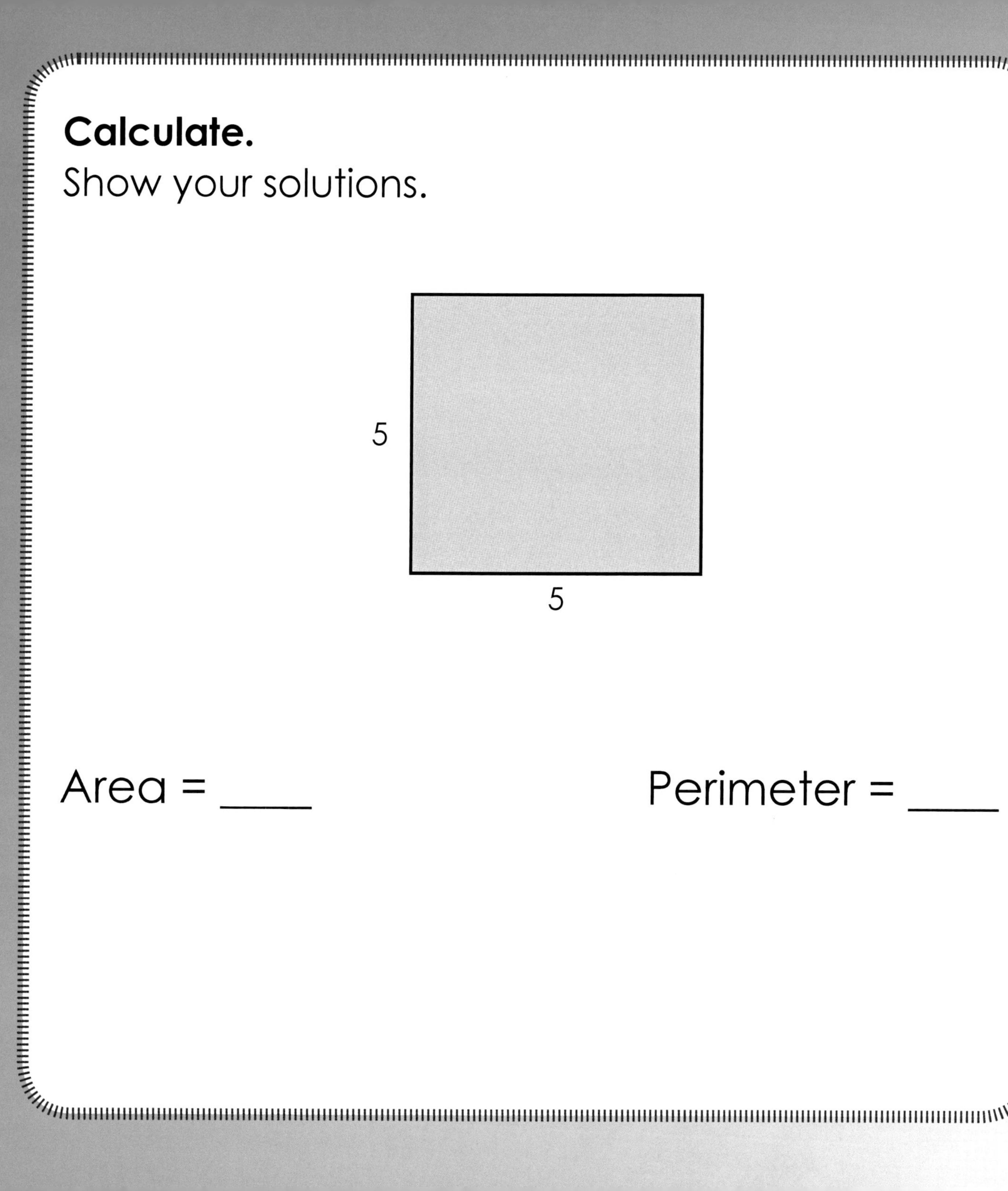

Area = ____

Perimeter = ____

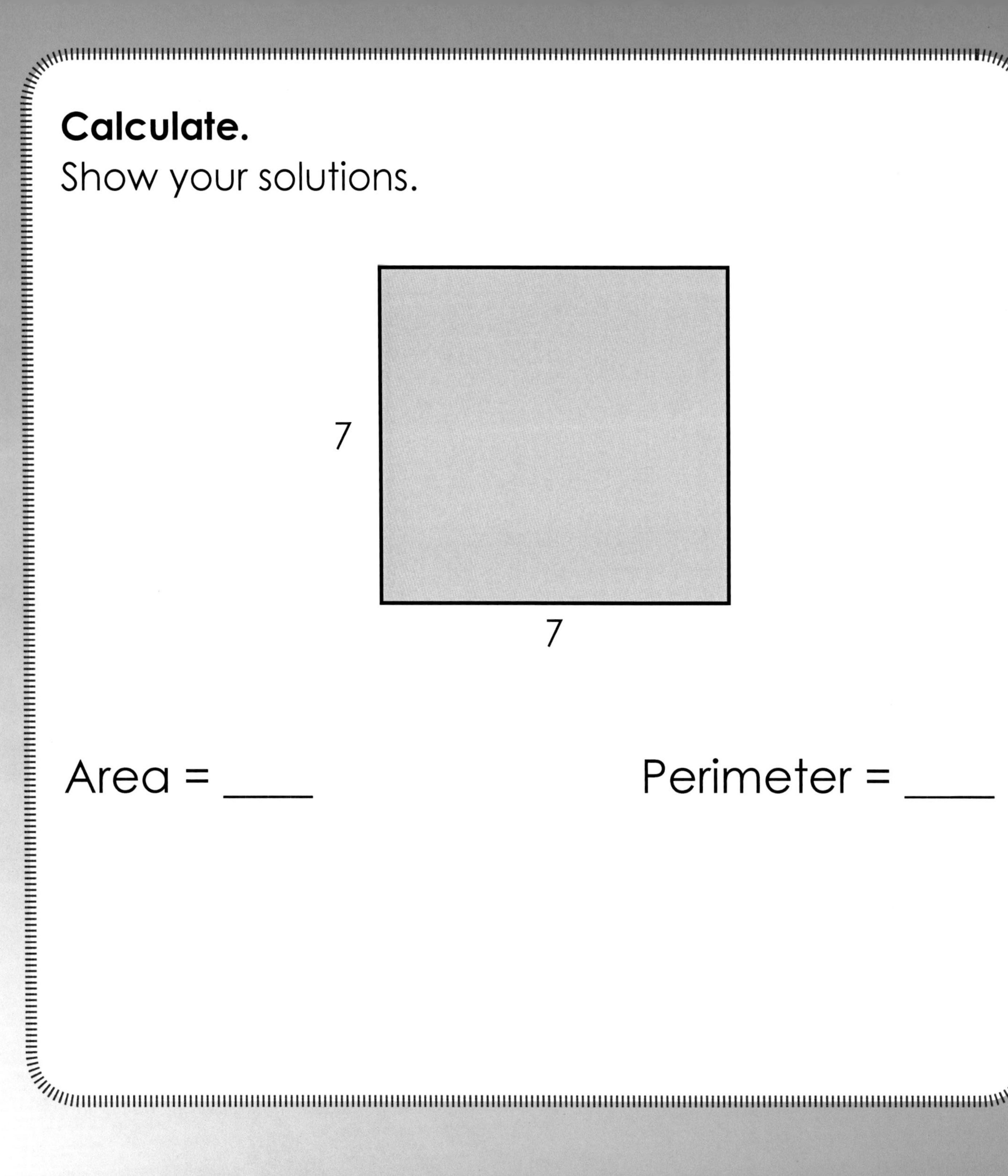

Calculate.

Show your solutions.

Area = ____

Perimeter = ____

Introduction to VOLUME

The surface area of this cube is 96 square units.
What is its volume?

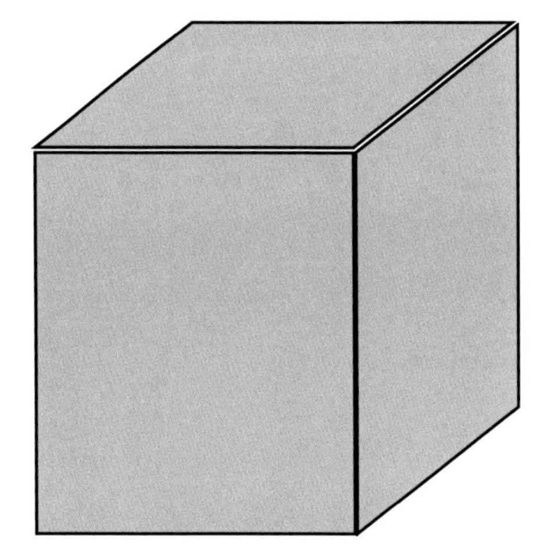

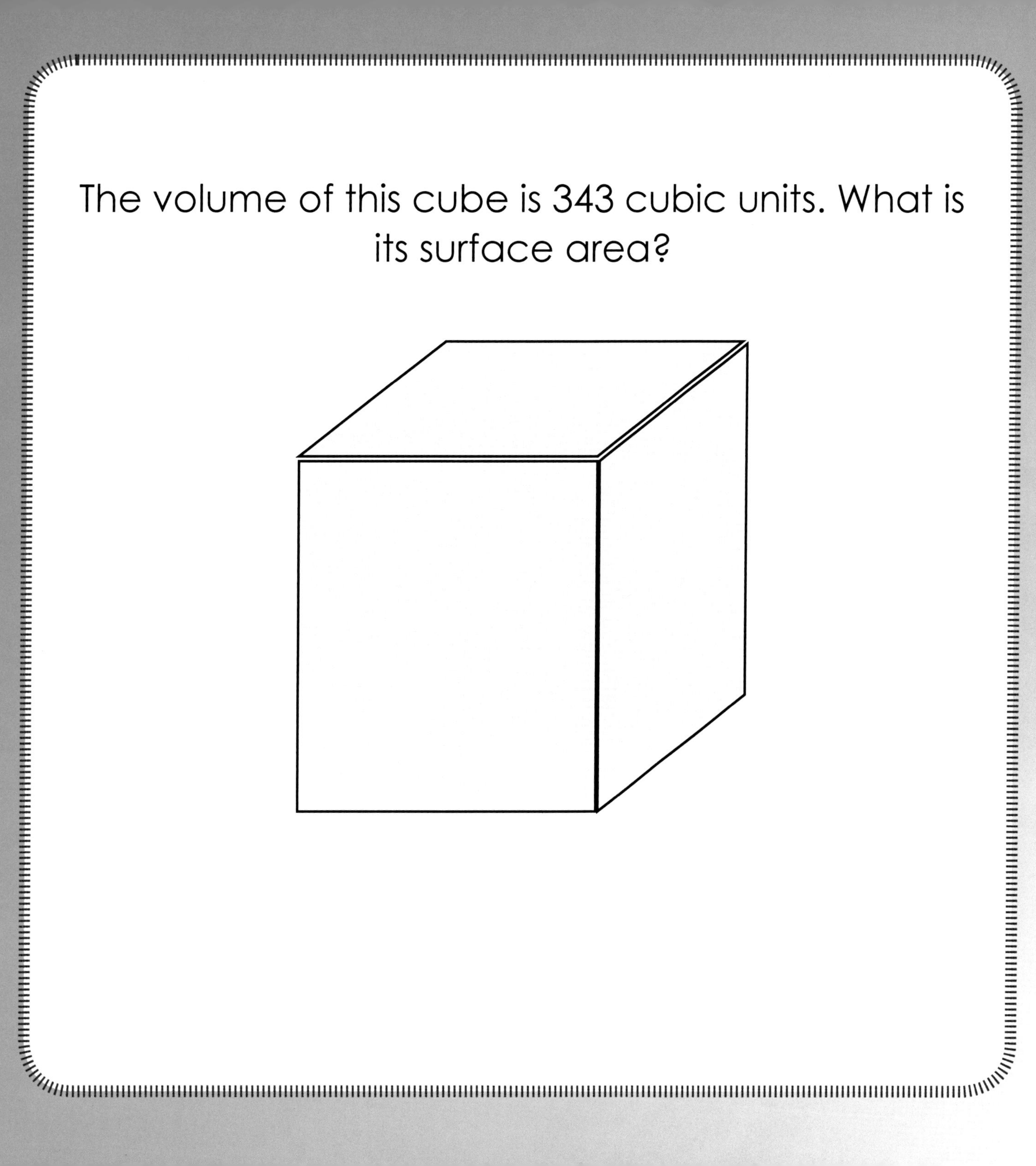
The volume of this cube is 343 cubic units. What is its surface area?

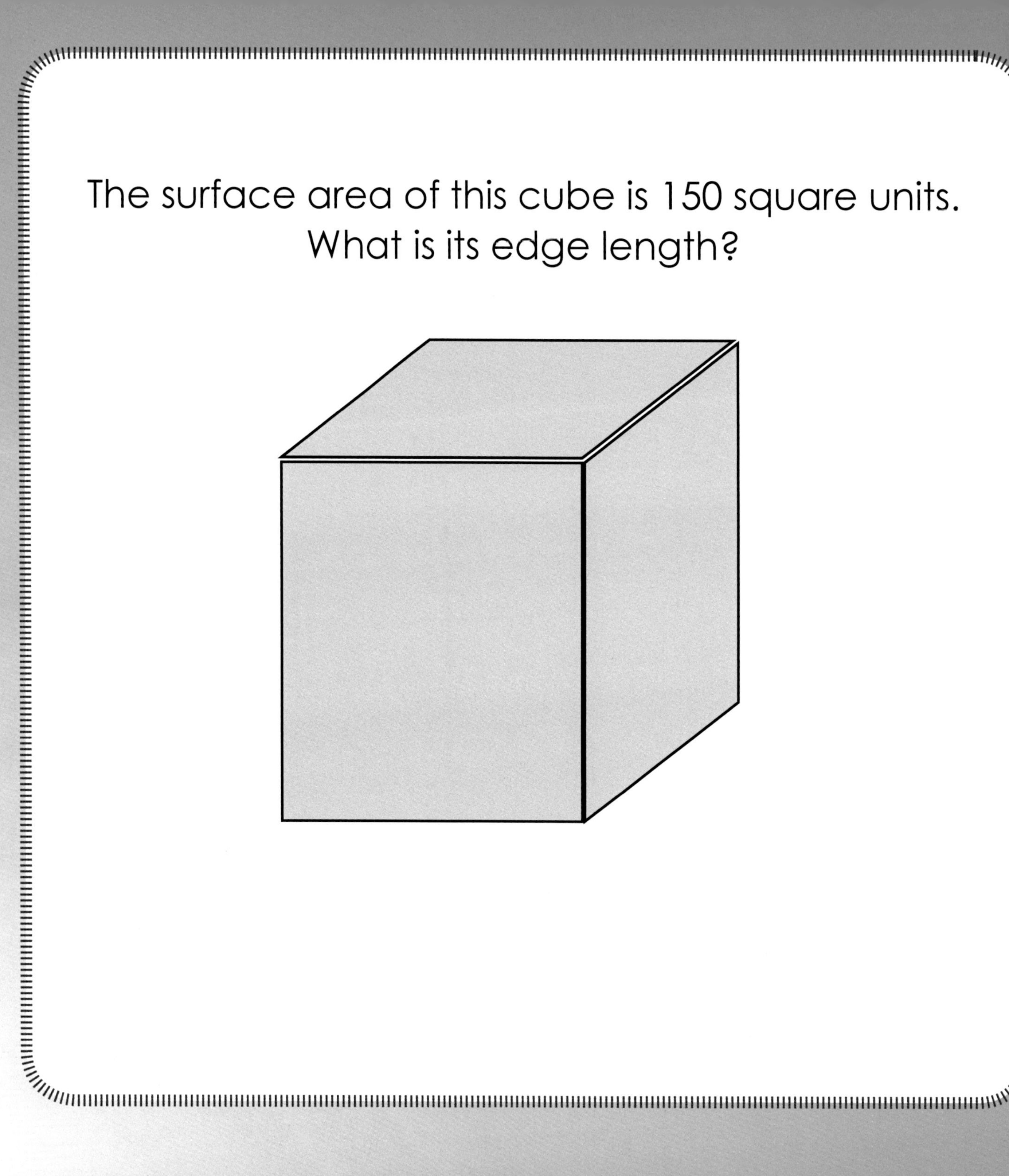
The surface area of this cube is 150 square units.
What is its edge length?

Answer

Triangles

1. acute
2. obtuse
3. right
4. obtuse
5. right
6. right

1. A = 30; P = 25
2. A = 15; P = 21
3. A = 8; P = 12
4. A = 12; P = 17
5. A = 20; P = 21
6. A = 9; P = 13
7. A = 18; P = 18

Circles

1. C = 18.85; D = 6
2. C = 31.42; D = 10
3. C = 43.98; D = 14
4. C = 56.55; D = 18
5. C = 25.13; D = 8
6. C = 37.7; D = 12
7. C = 50.27; D = 16

Squares/Rectangles

1. A = 32; P = 24
2. A = 30; P = 22
3. A = 24; P = 22
4. A = 25; P = 20
5. A = 49; P = 28

Volume

1. The volume is 64 cubic units.
2. The surface area is 294 square units.
3. The edge length is 5 units.

Made in the USA
Columbia, SC
07 July 2023